Life is not possible without the
Property
Freedom is not possible witho

.Building Durable Freedom

Compiled and written by Don and Linda Hansen

A Prologue

This book will show you how to achieve a beautiful and elegant societal culture.

Freedom is when you have 100% control of your property 100% of the time. Property and Freedom are connected; without property there is no freedom.

The Property Principle is the answer to every single human dilemma in life, including the work of Building Durable Freedom. The Property Principle is the answer to every question of right and wrong, every disagreement, every success, every disaster in life is easily reconciled with a single simple question:

Whose property is it?

The simply profound answer: If it isn't yours, leave it alone!

You need go no further in resolving any issue. Test it out! The indisputable theory of the Property Principle works 100% of the time. Learn to work with it. In time, you'll find that the quality of human culture is measured by how much positive productive property is respected.

Property is your life and all that you build using your life. Property is just three things: Ideas, the results of your ideas and thinking, and your own cellular structure.

Life is not possible without the defining boundary condition of Property.
Freedom is not possible without Property protection.

Using the simple Property Principle, you'll find that Political Law books are not needed, just as political arguments and wars are not needed. Just the Property Principle is needed to guide mankind's intellect and goals, and to guide mankind "this way to the stars".

Allow yourself to be open to education, which is neither an economic nor a class factor. It is a simple a matter of curiosity about the truth using to the Property Principle.

In today's world, more truthful communication and fundamental rightness is key.

An example of what's wrong in human culture is today's "politically correct" educational system that is busy pouring out (and passing on) high school and college graduates who are neither able to speak a single grammatically correct sentence, nor able to articulate an original thought or idea. Today's educators are ignoring truthful history, changing history, & deleting history, with the false belief that they are somehow influencing or saving today's students from the historical truth about themselves, their pasts, and their futures. This wrong process is doomed to fail. History is what it is... an absolute irrefutable fact of past occurrences. To lie about our past history is a huge violation of the Property Principle.

Please read on ...

Building Durable Freedom

Compiled and written by Don and Linda Hansen

FOREWORD

To the Book "Building Durable Freedom".

Close your thoughts to the day, and your eyes. Listen, listen for their cries, their minds urging forward no more.

Life is not possible without the defining boundary condition of Property.
Freedom is not possible without Property protection.

Think of the millions whose lives were stolen, enslaved, murdered, used, and thrown carelessly aside by thieves of the soul, by horrible ruthless men who thirst for control over other men.

This book will touch you with the great truth that human Freedom is all around you now, and has always been around you.

Which way do you turn toward Freedom? Which path, which step, which direction, starts the process … the sure asymptote[1] toward Freedom?

For most everyone, the path and the step are inborn, ready for propulsion onward. It is your choice to proceed. Do so now! Turn the following pages slowly, absorbing the simple concepts as tasting familiar freshness maybe for the first time.

Everything written here is the truth, with waypoints, and complete enough to let mankind, at last, turn toward and build a durable Freedom for all living beings.

Prerequisites for absorbing and understanding this book are an interest in finding the truth, and later an ability to consistently think in rational terms of observable, valid, truthful principles. You will find tools in this book to scrape away preconceived "truths" and replace them with valid truths. The basis of what is true and valid is found using the reliable, self-correcting, rational truth machine … the Scientific Method[2]:

---

[1] Asymptote, a value that you can get ever closer to but never quite reach.
[2] Andrew J. Galambos, *Sic Itur Ad Astra*, The Universal Scientific Publishing Company, San Diego, California, Page 65. This book presents the foundation of Volitional Science.

Life is not possible without the defining boundary condition of Property.
Freedom is not possible without Property protection.

The Scientific Method

1. **Observation** (for data gathering, ie: wars are a detriment to living organisms)
2. **Hypothesis Formulation** (only politicians start wars).
3. **Extrapolation** (Extend the hypothesis, ie: all wars enrich banks & politicians)
4. **Observation Again** (for corroboration) – Testing. Avoid politics to avoid war.

\+ Use **Occam's Razor**[3] for fewest unproven assertions.

---

[3] After William of Occam (1287 – 1347)

Life is not possible without the defining boundary condition of Property.
Freedom is not possible without Property protection.

Chapter One

Life is not possible without the boundary condition of Property:

Property is defined as[4]: A volitional being's individual Life and all non-procreative derivatives of that Life.

A volitional being is defined as: A lifeform able to identify and remember desires. A volitional act is an act of choosing.

Procreative derivatives, our children[5], are not our property. They arrive needing only the respectful nurturing and guidance of loving human beings, but their struggle for independence, and their inborn need to be volitional beings, free to choose, free to independently make choices for their lives began to further evolve at first breath.

Most, maybe all, of society's problems, from theft to murder to politics, have one common thread. Amongst the thieves, murderers & crooked politicians, none have a sense of property and its identifying boundary…they use "what's yours is ours to control" mentality. These thugs cannot produce positive value for themselves.

A seemingly simple, singular, primary, all-important concept of "The Property Principle", if implemented in a world ready for the solution it offers, could profoundly change human lives, and build durable freedom.

---

[4] Andrew J. Galambos, *Sic Itur Ad Astra*, The Universal Scientific Publishing Company, San Diego, California, Pages 21-23.
[5] Genetic Publishment of a new separate individual that occurs at conception from discarded egg and discarded sperm. So, this new individual with its own property rights becomes a temporary parasite within the mother.

Life is not possible without the defining boundary condition of Property.
Freedom is not possible without Property protection.

The Property Principle: If something isn't yours, leave it alone,
except by way of explicit permission !

Start using the Property Principle around the house, at work, in agreements, etc., and watch a stabilized, durable, and positive productive culture grow. There is a simple solution to deal with property violators – See and avoid them, and honestly advertise their wrongdoing. Realize that property violators need you far more than you need them.

Property has three components[4]:

1. **Primary Property** is the Volitional beings' thoughts and ideas. "These are electrical impulses arranged within the brain system, and they define the individual[6]."
2. **Secondary Property**, the volitional being's production that flows from Primary Property.
3. **Tertiary Property**, the volitional beings outermost natural boundary condition covering the body and its internal workings.

Energy is directed and exchang6ed across Property boundaries. Life is unique in that Life revises the form and productive/reproductive utilization of stored energy within its Property boundary[7]. Directed energy is thus represented as a vector quantity, having the properties of magnitude and direction.

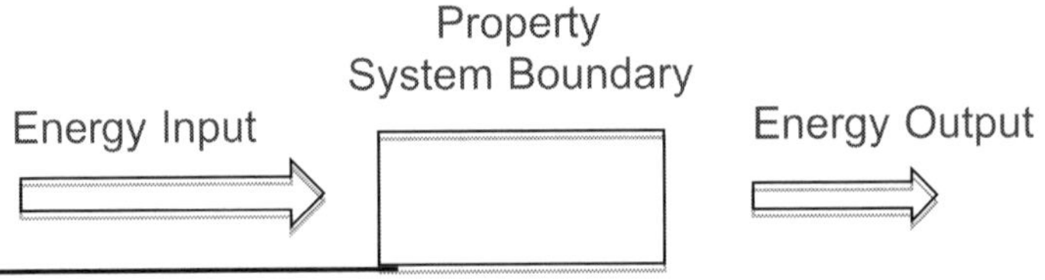

Property
System Boundary

Energy Input → [ ] → Energy Output

[6] Definition and characteristics of Life by Author: D. Hansen. Personal notes dated 1987.
[7] Definition and characteristics of Life by Author: D. Hansen. Personal notes dated 1987. Life will be found all over the Universe as life is part of the universal workings machine.

Life is not possible without the defining boundary condition of Property.
Freedom is not possible without Property protection.

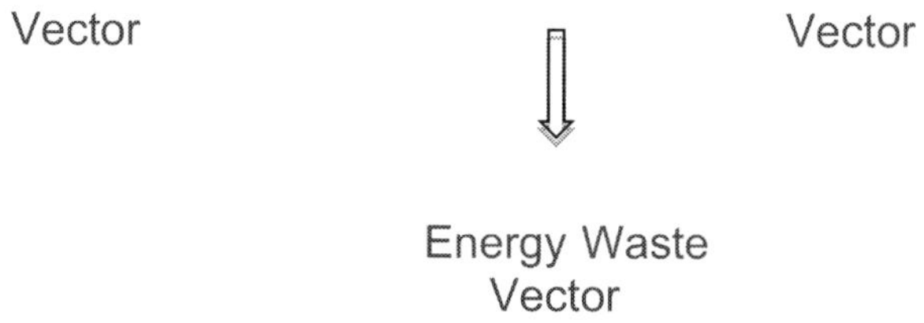

Vector ⇩ Vector

Energy Waste
Vector

This process diagram is common to all heat engines such as the human body, automobile engines, living single cell organisms, flowers, etc. All take in energy, utilize energy for output production, and all have waste products. Access to energy is access to Life.

From the above process diagram, a Second Law thermodynamic axiom for Life becomes apparent: All lifeforms use energy to create and maintain order in the Universe, order that counters the observable running-down of the Universe. Life's rearranged and maintained order thereby accelerates the running-down of the Universe, as entropy's Waste Vector is always present[8].

Lifeforms that obtain Energy Inputs from inanimate non-living nature (the Producers) are different from those that obtain Energy Inputs from other lifeforms through the use of theft (the ruthless Parasites and Predators).

Life creates and maintains its durable order both inside and outside its Property Boundary through the use of this Ideological Program[9]:

---

[8] Entropy: From the science of thermodynamics, the Second Law observation and measure that Energy use goes from order to disorder, from high energy to lower energy levels, i.e., there is no 100% efficient use of energy.

[9] Andrew J. Galambos, *Sic Itur Ad Astra*, The Universal Scientific Publishing Company, San Diego, California, Pages 340-342

Life is not possible without the defining boundary condition of Property.
Freedom is not possible without Property protection.

Innovation, followed by **Education**, **Advertising**, and **Maintenance**.

> **Innovation** is the process of discovery of natural Laws, and the inventions that come from the application of those Laws.
> **Education** is the transference of Innovation knowledge to others.
> **Advertising** is image building and further knowledge transference.
> **Maintenance** is preservation of resultant production.

Of the above four flow stream elements, Innovation is the most important and thus the most valuable. Education, Advertising, and Maintenance are very important, but of progressively lower importance than Innovation.

> Importance is defined as the amount of Property affected[10].
>
> Value is defined as a contractual or trade agreement, which is a mutual value disagreement that makes for a successful marketplace trade[11]: In other words, what value you obtain during a trade is more valuable than the price you paid, and, the price received by the seller from you is more valuable than the seller's value traded.
>
> Price is defined as an agreement for a current value, an accounting system currency.

To maintain Property, a lifeform must be free to live ... which means to freely gather, utilize, and store energy. The Concept of Freedom is presented in the next chapter. Freedom, Property protection, and Liberty operate together to build lasting Freedom. Liberty is your Freedom to act with your

---

[10] Andrew J. Galambos, *Sic Itur Ad Astra*, The Universal Scientific Publishing Company, San Diego, California, Page 289
[11] Andrew J. Galambos, *Sic Itur Ad Astra*, The Universal Scientific Publishing Company, San Diego, California, Page 203

Life is not possible without the defining boundary condition of Property.
Freedom is not possible without Property protection.

Property as you wish[12], so long as your actions do not interfere with the Property of others.

Eighteenth century American thinker John Adams observed that Property must be preserved or Liberty cannot exist, as did his contemporary colleague George Washington who declared that private Property and Freedom are inseparable.

---

[12] Andrew J. Galambos, *Sic Itur Ad Astra*, The Universal Scientific Publishing Company, San Diego, California, Page 22

Life is not possible without the defining boundary condition of Property.
Freedom is not possible without Property protection.

Chapter Two

<u>What is Freedom?</u> Freedom is the societal condition that exists when every individual [volitional lifeform] has full (i.e. 100%) control over his own Property[13]

> *"Freedom is not an empty sound; it is not an abstract idea; it is not a thing that nobody can feel. It means, - and it means nothing else, - the full and quiet enjoyment of your own property. If you have not this, if this be not well secured to you, you may call yourself what you will, but you are a slave"* [14].

If we humans have souls, it is most probably life's natural imperative to rationally create order within the Universe. Downstream of this imperative is the conversion & exchange of energy across the boundaries of each lifeform, as below:

Energy in … Energy Out … Energy as Waste Products (see page 2 for diagram and definition of Vectors): Any interference with a lifeform's system boundary, or any of the individual lifeform's created energy vectors, is a loss of Property or Property potential, and thus a loss of Freedom.

Protecting Property is essential to creating and maintaining Freedom. Protecting Property is the energy-using technological mechanism that builds and maintains the condition of Freedom: Knowledge accumulation and rationality are required.

---

[13] Andrew J. Galambos, *Sic Itur Ad Astra*, The Universal Scientific Publishing Company, San Diego, California, Page 25. [modification added by authors].
[14] William Cobbett (1763-1835). Even this English Parliamentarian knew Freedom's truth, but apparently did not see that Freedom is unachievable within the coercive societal theft mechanism organization called Politics.

Life is not possible without the defining boundary condition of Property.
Freedom is not possible without Property protection.

Lifeforms that cannot produce and maintain Property are not capable of achieving their own Freedom: Knowledge accumulation and rationality are required.

Property is arranged and maintained by rational volitional beings within their own created boundary condition, a condition that is an honest, reliable, and durable accounting for positive productive value.

Freedom then, naturally allows for the production of a rational Culture, a societal condition wherein Property is protected, insured, and restituted easily. Simply speaking, a societal Culture is when and where worthwhile (rational)[15] beings get together to do worthwhile (moral)[16] things in a worthwhile place, and where apology + restitution is the pathway out of difficulty.

Freedom is most often destroyed when a worthwhile Culture is interrupted by irrational beings that cannot and/or will not choose rationality over irrationality to determine ownership of Property. Irrationality and immorality is always natural for these destructive lifeforms. Their failure to use the Scientific Method is the largest tip-off to discover the magnitude and direction of future negative events.

There will always be a natural effort to interrupt the formation and maintenance of Property: An atmospheric storm, earthquake, taxation, a robbery, and many many other potentials. "The price of Freedom is eternal vigilance"[17]

---

[15] Rational means adhering to truthful validity (The Scientific Method)

[16] Moral means the absence of coercion. Coercion: "The intended intentional interference with the Property of another". Andrew J. Galambos, *Sic Itur Ad Astra*, The Universal Scientific Publishing Company, San Diego, California, Page 23.
[17] Thomas Jefferson, (1743 – 1826)

Life is not possible without the defining boundary condition of Property.
Freedom is not possible without Property protection.

Freedom is a product to be invented, built, and maintained to form a guiding structure for volitional individuals to live and to trade with one another across their boundary conditions defined by Property. Property and its boundary is a stable structure that can change with circumstance, but is firmly founded upon first principles of a science (truth[18]), and the rational application of the Scientific Method.

Building basic Freedom requires some wealth, some knowledge of Freedom, and some ability to see and avoid theft & destruction.

Why has there been so much destruction of Freedom? Because ruthless Predators and Parasites need to control the values that they can't produce. Their resultant property destruction is identified by way of the tattered, pierced, and destroyed boundaries of the individual's Freedom.

It is much easier to steal or destroy value than it is to create value.

"Sleeping nations either die or wakeup as slaves" Kemal Atatürk (1881-1938)

The human sociopathic/psychopathic brain types, "The ruthless versus the rest of us"[19], mostly hide the truth behind their creation of homogenized cultural confusion. Their intention is destruction of the individual's Property boundary through the use of their purposeful control methodology of cultural homogenization and homogenization of all other

---

[18] "I would say that education is meant to give us an idea of truth", Booker T. Washington

[19] Populated by ruthless predators and parasites in every lifeform. See "The Sociopath Next Door" by Martha Stout, 2005, Random House, New York, a very good first primer on sociopathy/psychopathy. Sociopaths fake most everything to gain advantage over unsuspecting empathetic normal people.

Life is not possible without the defining boundary condition of Property.
Freedom is not possible without Property protection.

things, outside themselves. Such destruction is necessary for them to more easily control the world they dream of...their false make-believe world is built upon constant lies and other coercion[20]. This homogenized sociopathic world coercively controls the bothersome "common eaters" that are placed far below the status of the sociopath's self-convinced pompous superiority - a false superiority that ignores the productive human nature of yearning for Freedom plus the secure boundary conditions of Property.

Once humankind's ruthless ones have you convinced to accept their homogenized one-world, your individualist diversity will not be tolerated by them. Their centralized control mechanism usually condenses into some coercive Communist/Fascist political state, or some other coercive force and fraud centralized control organization with tribal chief and witchdoctor functionaries.

A sociopathic swarm is naturally formed with their like-minded brethren. Their grouping method is intended to confuse and overwhelm all opposition with a cacophony of coercion, similar to a swarm of stinging insects.

Individuals and Individualism become crimes against their political states.

Fortunately, such group homogenization is centralized in concept and practice, a condition that ignores the individual's human nature of decentralization of control mechanisms. Coercive control centralizations are always on a heading for destruction, and this unproductive characteristic is the tipoff for their eventual downfall.

---

[20] Coercion: "The intended intentional interference with the Property of another". Andrew J. Galambos, *Sic Itur Ad Astra*, The Universal Scientific Publishing Company, San Diego, California, Page 23.

Life is not possible without the defining boundary condition of Property.
Freedom is not possible without Property protection.

Individualism's alternative productive decentralized trading systems are inherently stable due to Energy Vectors having many load/stress sharing redundant structures and pathways that operate to smooth out foundational and operating perturbations[21]: Insurance, or risk purchasing, is one such productive system, voluntary division of labor and decentralized creation of Money[22] are other ones.

The human sociopathic/psychopathic brain types (ruthless predators and parasites) start with small Property violations, increasing their violations to the point of a pushy confusing intrusion that finally becomes a forcible injection across property boundaries.

Eventually, their intrusions are seen by most of the controlled humans to be against productive human nature, at which time a rebellion ensues.

Destructive sociopathic/psychopathic brain type characteristics are inborn and are unlikely to be changed. So, their final coercive central control mechanism is copied, implemented, and maintained across the ages in a robotic fashion. The colors, flags, institutions, slogans, etc., are the only variances. In the end, their plans always proceed through the use of murder and intrigue.

These robberies of Life are the prime reason that politics is one of mankind's most serious problems: Politics and political states.

> "This is the tendency of all human [political] governments. A departure from principle [property

[21] Perturbation, small changes in value to redirect a process more toward validity.
[22] Definition of Freedom's Money: A durable & reliable accounting system for positive productive value. Defined by by author D. Hansen 2015, August, using capital "M" to denote real Money.

Life is not possible without the defining boundary condition of Property.
Freedom is not possible without Property protection.

> principle] becomes a precedent for a second; that second for a third; and so on, till the bulk of society is reduced to mere automatons of misery, to have no sensibilities left but for sinning and suffering... And the fore horse of this frightful team is public debt. Taxation follows that, and in its train wretchedness and oppression." -- Thomas Jefferson [emphasis added].

> "Government is best which governs not at all; and when men are prepared for it, that will be the kind of government they will have." -- Henry David Thoreau

> "The easiest way to gain control of the population is to carry out acts of terror." The public "will clamor for such laws if their personal security is threatened." -- Joseph Stalin

Thoreau's lesson is basically…honestly govern yourself, and become part of an honest trading society that is populated by other honest self-governing individuals.

Stalin's lesson produces Theftism[23], which is a descriptive concept for the actions and organizations assembled by the sociopathic/psychopathic brain types for the purposes of parasitism, theft, murder and intrigue. The term Theftist or Theftism replaces communism, socialism, crony-capitalism or mercantilism, fascism, kings & queens, dictators, murderers, crime bosses, petty thieves, robbers, taxation, political office holders, etc., etc.

The term Theftism gathers all property violation potentials under one heading that nails down the operative truth, which is the corrupt actions and intents of the "thieves of the soul": Theft.

Theftists dearly love making big plans to control the lives of others, and always at the expense of those "others".

---

[23] Defined by Author D. Hansen 2011, October

Life is not possible without the defining boundary condition of Property.
Freedom is not possible without Property protection.

You might be able to calculate your Freedom ratio or percentage this way[24]:

$$\% \text{ Freedom} = \frac{(\text{Your Property} - \text{tax, other theft, etc.})}{(\text{Your Property})} \times 100$$

[24] In today's world, this % is always far less than 100%. The positive productive goal is to increase Freedom to approach 100%.

Life is not possible without the defining boundary condition of Property.
Freedom is not possible without Property protection.

Chapter Three

The Nature and Significance of Good and Evil. Good is defined as: A choice which is Rational plus Moral[25]. Therefore, evil is defined as a choice which is irrational and/or immoral.

Rational means adhering to truthful validity (The Scientific Method), and Moral means the absence of coercion. This is the identifying nature of Good and Evil.

The significance of Good is using energy to produce stabilized, moral, rational, and durable positive productive property value. The significance of Evil is theft and destruction of property.

It is important to realize something very natural in humans and most all other life capable of rational thinking: The producer life form versus the ruthless predator/parasite lifeform is a clash of energy gatherers/users.

The naturalness of the above sentence is likely best identified by the differences between brain structures suited for long term production, and brain structures with certain parts left out that render them short term operatives. Short term thinkers are unable to optimize durable production and productive behavior over the long term or sometimes even their own short term attention spans.

So, the natural clash between the productive brain types and the ruthless sociopathic/psychopathic parasite/predator brain types, or good versus evil, is rooted in how energy is obtained.

---

[25] Andrew J. Galambos, *Sic Itur Ad Astra*, The Universal Scientific Publishing Company, San Diego, California, Pages 74, 93

Life is not possible without the defining boundary condition of Property.
Freedom is not possible without Property protection.

Major Producers think, act, take risks, and obtain energy from inanimate non-living lifeforms consistently over the long term species timescale. So, the clash can also be seen as between the long term species timescale versus the short term range-of-the-moment time scale … the rational and moral versus the irrational self-interested scheme-defined timescale of the sociopathic/psychopathic parasite/predator

"Minds differ still more than faces" Voltaire (1694-1778)

Below are some written examples of the always-predictable, irrationally self-interested non-empathetic sociopathic/psychopathic parasite/predator brain types, the ones that use murder and intrigue to destroy, then rebuild, a world "to our hearts desire" [26]**:**

First, from the books "Rules for Radicals, Reveille for Radicals"

> by Communist[27] writer Saul David Alinsky (1909 -1972), busy scheming on the subject of how to create a Socialist[28] political state using coercion to control the producers. In other words, how to institute Theftism:
>
> Alinsky declares eight prime levels of coercive control that must be obtained before you are able to create a socialist political state:
> 1) Healthcare – Control health and you control life and the fear of losing life.

---

[26] Wolves in Sheep's Clothing, The Fabian Society in England, 1884, is typical of sociopaths/psychopaths. Study the Fabian socialists to see and avoid these types of predators.
[27] Communism, the international application of Theftist socialism for the entire world, and later the Universe, to be administered *uber alles* by predators. Fascism is National Socialism.
[28] Socialist, Socialism. A form of political state Theftism.

Life is not possible without the defining boundary condition of Property.
Freedom is not possible without Property protection.

2) Poverty – Increase poverty because poorer people are easier to control.
3) Debt – Add to all forms of debt: Personal, business, and public political state debt. Increase taxes[29] to increase debt ever more.
4) End the American Second Amendment right to bear arms – Defense is always moral, so removing the ability to defend against tyranny weakens the victims of Theftism.
5) Tax Supported Welfare – Forcible Theft from some to support others results in controlling the means to produce property.
6) Education – Tell lots of lies that seemingly support the socialist political state.
7) Philosophy/Religion – Remove the belief in anything greater than the coercive Theftist political state.
8) Class Warfare – Foster groups as opposed to individuals, and then turn the groups onto each other. Tell more lies, excuse wrongness on "our side", then get a fight started. Tyrants love class warfare. They love even more stepping into any fight to take control of the warring groups these tyrants initially provoked. They call this operation "bottom up (begging for relief), top down (them) to quell dissent, then emotional inputs to make you glad you have their brand of "peace".

Alinsky copied Lenin's scheme for one-world-homogenized society conquest using political communism, under Russian rule. Stalin described his converts as "Useful Idiots." Beware of anyone's Useful Idiots.

---

[29] Taxes, taxation, tax: Forced payment out of productive wealth from producers to benefit some political state or other crime organization doing a monopoly service that you may not want or need. Payment is extracted at the point of a gun or through the use of some other force and fraud mechanism. Without taxation and dishonest banking, force and fraud organizations have no wealth because they do not know how to produce positive productive value.

Life is not possible without the defining boundary condition of Property.
Freedom is not possible without Property protection.

"It is difficult to free fools from the chains they revere."
Voltaire (1694-1778)

There is only one way to create a societal structure like those of Lenin and Alinsky: Through the use of coercion, or force and fraud, to destroy the individual and his property for use by the centralized group controllers.

The success of Theftist force and fraud requires the wholesale theft of wealth and telling lots of lies (especially with money) to individuals who are unprepared with a rational and moral philosophy to guide their thought processes.

Theftists attempt to divide cultures into opposing political groups: For example, the poor groups are set against the rich in an effort to rob the rich, citing the falsehood of the static "property pie". Meaning that everyone must fight another person for a "piece of the pie". After the intended and induced mayhem matures, the guilty scheming sociopaths descend into the fight to calm everyone down into political servitude[30].

The above paragraph illustrates that Theftism's success is in corralling a majority of individuals into some kind of political organization to be sure that this majority has accepted the concept of political control. This is Theftism's primary entrapment scheme … convincing individuals of the falsehood that politics or something similar is needed to produce harmony and peace.

---

[30] "*The Road to Serfdom*", 1944, by F. A. Hayek, the Austrian economist/philosopher.

Life is not possible without the defining boundary condition of Property.
Freedom is not possible without Property protection.

Second, from mankind's unhappy non-friend Ian L. McHarg (1920-2001) a book:

"*Design with Nature*". McHarg wistfully writes hypothetically that, after mankind destroys the Earth, there appears from within a deep-laden slit a lifeform peeking out and declaring: "Next time no brains", a grand irony that this new lifeform had the "brains" enough to see how awful is McHarg's hypothetical effect wrought by mankind. By blaming all mankind for some of the Earth's real or imagined scars, sham environmentalism sets up a pulpit of camouflage for the mankind-fearing sociopaths to push for their ruler-ship "*uber alles*", a political dictatorship that controls life and life's positive production while implementing the swindle of "saving" mother earth. Again, dividing individuals into warring political groups.

Third, the book "*Tragedy and Hope*" by Carroll Quigley:

"*The powers of financial capitalism*[31] *had another far-reaching aim, nothing less than to create a world system of financial control in private hands able to dominate the political system of each country and the economy of the world as a whole. This system was to be controlled in a feudalist fashion by the central banks of the world acting in concert, by secret agreements arrived at in frequent private meetings and conferences. The apex of the system was to be the Bank for International Settlements in Basel, Switzerland, a private bank owned and controlled by the world's central banks which were themselves private corporations. Each central bank ... sought to dominate its government by its ability to control Treasury loans, to manipulate foreign exchanges, to influence the level of economic activity in the country, and to influence cooperative*

---

[31] Or, crony-capitalist mercantilism which are Political state granted monopolies like Charles II's Hudson's Bay Company (1670), private banking's Federal Reserve Act of Congress (1913), etc.

politicians by subsequent economic rewards in the business world."

Lastly, the books:

"*The Prince*" by Niccolo Machiavelli (1532) that urges the use of "effective truth" in lieu of actual truth, in other words lies by the bushel full to hide actual truth behind political confusion. "*The Prince*" is a resume' of sorts for Niccolo to get a job with the addressed political potentate. Niccolo outlines numerous political state Theftism schemes and offers to implement them.

"*The 48 Laws of Power*", by Joost Elffers and Robert Greene, 1998, Penguin Books. This book is a well written expose' describing the ruthless dishonest control tactics and schemes of the sociopath/psychopath. Typical are narcissistic schemes such as "make yourself a magnet of attention … more colorful than the bland and timid masses". The observational data in this book could be used to create an artificial intelligence computer program to enable producers to see and avoid sociopaths and psychopaths. After all, the predators are very busy using similar computational techniques, cameras, eye scanners, etc., to identify and censor those who like freedom.

The sociopaths and psychopaths can be charming, endearing, and/or a group of pushy, demanding, and militant noise makers comprised of people unable and/or unwilling to do critical thinking to find truth in their surroundings. They see themselves as victims of the effects of Freedom. They are overwhelmed with hatred and jealousy that there are productive people who are everything that they are not.

"... Producing a kind of painless concentration camp for entire societies so that people will in fact

Life is not possible without the defining boundary condition of Property.
Freedom is not possible without Property protection.

have their liberties taken away from them, but will rather enjoy it, because they will be distracted from any desire to rebel by propaganda, or brainwashing, or brainwashing enhanced by pharmacological methods. And this seems to be the final revolution." -- Aldous Leonard Huxley (1894-1963)

Now, for some written examples of positive productive work done by rational and moral producers. These thinkers show that the political state is not needed for humankind and other likeminded lifeforms to advance into the future:

"*The Mainspring of Human Progress*" by Henry Grady Weaver. Tells the story of positive productive progress made by mankind throughout history.

"*The Road to Serfdom*", 1944, by F. A. Hayek, the Austrian economist/philosopher warns of the tyrannical central planning takeover of marketplaces.

"Atlas Shrugged", by Ayn Rand, 1957. A novel that accurately represents the clash between producers and the predator/parasites. All the other works of Rand are very worthwhile reading, starting with "*Anthem*".

"Thomas Paine Author of the Declaration of Independence" (1947), and *"The Tragic Patriot*" [32] (1954), both by Joseph Lewis. The American "*Declaration of Independence*" (1776) can be used as a template for your own improved declaration of Independence from Theftism.

---

[32] Patriotism, not a political concept, is one's liking, respect, and acknowledgement for values received during one's upbringing.

Life is not possible without the defining boundary condition of Property.
Freedom is not possible without Property protection.

"The Discovery of Freedom" (1943), by Rose Wilder Lane, Describes Man's struggle against authority.

"Economics in One Lesson", by Henry Hazlitt (1946) explains the fallacies of socialism (Theftism) as presently practiced worldwide.

"Lightning in His Hand the Life Story of Nikola Tesla", by Hunt and Draper (1977) and "Prodigal Genius the Life of Nikola Tesla" (1944), Story of a master innovator and discoverer of scientific principles. Like Newton, the world stands on the broad shoulders of these modern productive minds in ancient bodies. Freedom stands on these shoulders as well because the structures of Freedom are based upon scientific principles.

"*The Cry and the Covenant*" (1949) by Morton Thompson. The story of the discovery of contagion, illustrating that discovery using valid principles saves lives.

Andrew J. Galambos, "*Sic Itur Ad Astra"*, The Universal Scientific Publishing Company, San Diego, California. *"This way to the Stars"*, the basis for the Theory of Primary Property.

Life is not possible without the defining boundary condition of Property.
Freedom is not possible without Property protection.

Chapter Four

Money. "Trading Value for Value"[33]. A Trade is a barter between owners of property. There is a one-to-one correspondence between traded values. Money and its proper use make Freedom's stabilized durable growth possible. Presently, the commonly used monopolistic banking system (force and fraud system of money/currency) is one primary Theftist weapon that impedes the spontaneous outbreak of Freedom. If you'd like to know how human cultures sink lower with time, look to Theftism's systems of coercive monopolistic organizations as the Theftist pressure behind most all failures.

Money, currency, and banking are quite simple concepts: Real Money[34] is a durable, honest, and reliable accounting system for positive productive value. Money and its current value (currency) arise naturally within a durable, honest, positive, productive, and reliable marketplace. A Marketplace is where and when barter exchanges of value[35] are honestly and durably accomplished. Note that both Money and Marketplace have the same requirements, and that both require each other.

If barter of production becomes cumbersome, then another form of barter is introduced as a trade facilitator: This is a barter Money with agreed upon current value, or currency, set by the marketplace of traders. This facilitating currency is something desired by most traders in a given marketplace, a marketplace that assures accounting, reliability, honesty, and durability of the exchangeable currency.

---

[33] Ayn Rand, *"Atlas Shrugged"*, 1957
[34] Money defined by author D. Hansen 2015, August, using capital "M" to denote real Money: "A durable and reliable accounting system for positive productive value".
[35] For definition of Value, see page 4 above.

Life is not possible without the defining boundary condition of Property.
Freedom is not possible without Property protection.

In other words, the world runs on verifiable trust[36].

Real Money is production based. Un-real money is lie based, or partially lie based.

If the above is so simple, why then is the structure and operation of today's centrally and coercively controlled "money" system difficult to understand? The answer is simple but hidden from view behind the predator's curtain of purposeful confusion: Today's politically coerced "official" money/currency is a political state monopoly of uncertain backing that becomes a faulty loan at interest on your already created property. These "loans" to you, based upon your earned property value, are a theft coerced into use most often by some political force and fraud organization. Its currency value is decided by others and not the value-for-value trading partners. The needed one-to-one correspondence between traded values is disrupted to favor of the Theftist mechanism.

> *"I set to work to read the Act of Parliament by which the Bank of England was created. The investors knew what they were about. Their design was to mortgage by degrees the whole of the country ...lands ...houses ...property...labour. The scheme has produced what the world never saw before -- starvation in the midst of abundance."* -- William Cobbett (1763-1835)
>
> And:
> *"By this means government may secretly and unobserved, confiscate the wealth of the people, and not one man in a million will detect the theft." -- John Maynard Keynes (the father of 'Keynesian Economics' the Theftism scheme to which the World is now politically shackled) from his book "The Economic*

---

[36] Trust is defined and built by continuously keeping your word and performing on agreements.

Life is not possible without the defining boundary condition of Property.
Freedom is not possible without Property protection.

Consequences of the Peace" (1920), the "peace" after World War I, a war planned in the boardroom of banks to thrust central banking and other central controls onto the World.

On the unnecessary World War II: " *The war wasn't only about abolishing fascism, but to conquer sales markets. We could have, if we had intended so, prevented this war from breaking out without doing one shot, but we didn't want to." -- Winston Churchill to Truman (Fultun, USA March 1946)*

So, the fascism destroyed was the German/Japanese one, which made room for the British/American/Russian/etc. fascism.

Predators force the producers to allow these monopoly banks to declare the current value of the producer's productive work…thus causing a large scale coercive fraud, a theft because you and others must take interest bearing loans from these so-called banks to obtain politically approved official "currency" with which to trade: Thus creating a false debt based mortgage on the positive productive effort you've already created.

"If any man's money [Money = value of his labor] can be taken by a so-called government, without his own personal consent, all his other rights are taken with it; for with his money the government can, and will, hire soldiers to stand over him, compel him to submit to its arbitrary will, and kill him if he resists." -- Lysander Spooner (1808-1887) [Emphasis added by authors].

Lord Acton wrote in 1875 – "The issue which has swept down the centuries and which will have to be fought sooner or later is the people versus the banks." This is the same Lord Acton who stated: "All power tends to corrupt, and absolute power corrupts absolutely."

Life is not possible without the defining boundary condition of Property.
Freedom is not possible without Property protection.

Thomas Jefferson, letter to Samuel Kercheval, 1816 - "To preserve independence, we must not let our rulers load us with perpetual debt. We must make our election between economy and liberty, or profusion and servitude. If we run into such debts as that we must be taxed in our meat and in our drink, in our necessaries and our comforts, in our labors and our amusements, for our callings and our creeds, as the people of England are, our people, like them, must come to labor sixteen hours in the twenty-four, give the earnings of fifteen of these to the government for their debts and daily expenses, and the sixteenth being insufficient to afford us bread, we must live, as they now do, on oatmeal and potatoes, have no time to think, no means of calling the mismanagers to account, but be glad to obtain subsistence by hiring ourselves to rivet their chains on the necks of our fellow-sufferers." --

But of course, today's political rulers, who swear an oath to follow the rule of political Law seldom do follow these Laws that supposedly limit the theft of Property. Giving another person coercive dominion over your Life is permitting theft of all Property including your life as well: Thus introducing Theftism via consent.

An ideal culture will have no coercive rulers, just some productive marketplace leaders who earned their good reputations through the use of honest and productive dealings with others.

Solution: Manufacture a Money system and Value system that sees and avoids dishonesty. No longer accept the politically coerced money or currency. The "thing" that becomes a Money, as opposed to coerced lower case "money" (that isn't a real Money), can vary over time and circumstance. Money's longevity will be based upon honest, durable, and reliable accounting with storage of Value. The value is represented by what is traded, or that which is held by a trusted fiduciary[37] receipt for the value traded.

For thousands of years metals such as Gold have functioned as a reliable and desirable storage-of-value accounting system representing Money and currencies. Such metals are acceptable everywhere on earth even today and are mostly divorced from political state influence and interference. The prime reason metals are considered to represent Money and currency is: Metals are a receipt difficult to counterfeit and lie about. Long lasting metals set on a shelf tell the truth about stored Value, thus providing a good and durable accounting system for positive productive value.

> *"In the absence of the gold standard [meaning here a truthful, durable, and reliable standard], there is no way to protect savings from confiscation through inflation [the political state and coercive banking system telling lies with money]. There is no safe store of value. If there were, the government would have to make its holding illegal, as was done in the case of gold. If everyone decided, for example, to convert all his bank deposits to silver or copper or any other good, and thereafter declined to accept checks as payment for goods, bank deposits would lose their purchasing power and government-created bank credit would be worthless as a claim on goods. The financial policy of the welfare state requires that there be no way for the owners of wealth to protect themselves."*
> *"This is the shabby secret of the welfare statists' tirades against gold [and other Money forms]. Deficit spending [political state spending more money than it possess] is simply a scheme for the confiscation of wealth. Gold stands in the way of this insidious*

---

[37] Fiduciary: The holding of a trusted, durable, and ethical relationship that adheres to the positive productive rules of the marketplace and the traders of value-for-value.

Life is not possible without the defining boundary condition of Property.
Freedom is not possible without Property protection.

*process. It stands as a protector of property rights." -
- Alan Greenspan, "Gold and Economic Freedom" in
Ayn Rand, ed., Capitalism: The Unknown Ideal
(New York: Penguin Group, 1967), 101-108.
[Emphasis added by authors]*

Political state coerced money soon becomes a complete quagmire of theft**:** Bonds issued by politicians are "financed" by future tax receipts that are claims upon the productive lives of unsuspecting others. As fiduciary honesty decreases, these dishonest Bonds are increasingly financed by ever-larger lies about the quantity and quality of political money. Lies that are broadcast from behind the political curtain until the inflation[38] of political money becomes too obvious to ignore. The result is two bankruptcies: Fiduciary trust and their phony money/currency. Note that not one political money system survives very long. Collapse of these dishonest money systems is guaranteed.

"We can ignore reality, but we cannot ignore the consequences of ignoring reality." Ayn Rand, author of "*Atlas Shrugged*" (1957).

Mankind can invent Money and its currency through the use of many different trusted accountings that more nearly guarantee honesty: Electrical energy units, distributed accounting, metals, trusted Fiduciary receipts, etc.

A good place to start may be the rehabilitation and reintroduction of Adam's Smith's private Real Bills Doctrine[39] to finance world trade, driving out the dishonest politically coerced money put in place during the 1914 purposefully

---

[38] Monetary Inflation: Telling lies with money, made easier because real Money is driven out of the marketplace by dishonest money.
[39] *The Wealth of Nations* (1776). Adam's Smith's only error was his "time value" of labor, in lieu of production based value of labor.

caused World War I. War is a sociopath's mechanism to redefine rational society into their irrational heart's desire of homogenized centralized control of everything.

Life is not possible without the defining boundary condition of Property.
Freedom is not possible without Property protection.

Chapter Five

How to turn toward and begin building durable freedom.

It's easy in concept:

1. Keep your Agreements.
2. Protect & Respect all Property.
3. Use a Money system that provides an honest & durable accounting for positive productive value, and also provides a convenient value-for-value currency exchange mechanism.
4. Find methods to achieve a Recourse to Justice[40], usually through the use of good contract technology and insurance. Receipt of apology + restitution is always a Recourse to Justice.
5. Learn to see and avoid danger such as Theftism and other natural dangers. Build defensive mechanisms that preserve Freedom.

Harder in practice:

1. The Theftists have weapons and access to what you want to keep private.
2. They also cannot imagine losing what they have taken by force and fraud, so they expend great effort to maintain dominion *uber alles* (control over all persons).
3. However, paying for dominion *uber alles* comes from those productive souls under the boot-heel of the controlling sociopaths/psychopaths, a boot-heel that is always temporary.
4. See and avoid the boot.

Presently, the world is stumbling along with a claimed degree of smoothness. There isn't much to change for the start of

---

[40] "Justice [truthful correction of injustice] is the great work of humans on earth" – Daniel Webster [emphasis added]

Life is not possible without the defining boundary condition of Property.
Freedom is not possible without Property protection.

Freedom's beginning, however, limiting the theft of positive production is exceedingly important for Human life to continue:

1. Define the terms of your life, particularly define the words you use realizing that each word has an inventor and that word is the inventor's property (no redefinitions without explicit permission)[41].
2. Eliminate Politics to limit war and other thefts of production. Never give sociopaths/psychopaths a place to create fear. Know that your fear is just beyond the extent of your knowledge, so extend knowledge to evaporate fears.
3. Institute one-to-one correspondence with values and honest banking to limit theft of current values (honest Money). The amount of debt based dishonest money is presently huge. Sidestep this fraud by removing your wealth (false money, Money, tools-of-production, your home, protection technologies, etc.) away from the planned disaster of the Theftists.
4. Develop honest insurance mechanisms to limit the effects of productive losses.
5. Develop honest insurable contractual agreements that do not seek "legal enforcement" so that theft of production is minimized. The need to preserve reputation, one of our most valuable possessions, will generate an agreement keeping feature for all contractual associations.

Never again tolerate or organize a societal structure that includes coercion as a founding principle or uses coercion in any form. Instead, organize society so that property is protected and successfully defended.

---

[41] Semanticist S. I. Hayakawa's *How Dictionaries are Made* wrongly claims that usage determines definition. This wrongness becomes a Machiavellian "effective truth" after much misuse is accomplished, not recognizing that words have inventors and are thus primary property.

Life is not possible without the defining boundary condition of Property.
Freedom is not possible without Property protection.

Remember that Freedom is not possible without Property Protection.

Start with your own Declaration of Independence from Coercion, politics, force, fraud, Theftism, sociopaths/psychopaths, rude pushy people, etc. You don't need them, they need you.

Seek honest trading partners to enhance trade. Find prime personal contractual relationships for marriage, business, etc.

The Ideological Program from page 3 is a good structural outline with which to start building Freedom:

Innovation, followed by **Education**, **Advertising**, and **Maintenance**.

Innovate a culture with supporting societal structure.
Educate your trading partners to its nature and significance.
Advertise this new society to attract other like-minded individuals.
Maintain the society by defense of Property.

Preserve your wealth, preserve yourself.

Get out of debt.

Never trust a politician or their banker organizations, or any other unreliable third party to hold wealth. Trust a holder of wealth when insurable fiduciary trust is restored and honestly insured.

Simply turn toward Freedom. Rid yourself of predatory mental imprisonments thrust onto your human mind, such as the false idea that politics is needed to organize or build anything.

Life is not possible without the defining boundary condition of Property.
Freedom is not possible without Property protection.

Wage no war, no battle with sociopaths & psychopaths …just avoid these dangerous brain types and construct something new based on Freedom and defend this freedom-generating system. The sociopaths & psychopaths will bring coercion to you, so be able to defend your Freedom.

A moral and trustworthy insurance mechanism is a rational society's energy storage system for losses: A company that purchases your risk for a fee is one of mankind's best inventions.

On the subject of war:
War is always a theft operation of taking or destroying another's property. After the war, the expanded Theftist "banking" system is imposed on all sides for loans to rebuild what was purposely destroyed. Sidney George Reilly's (1873-1925) reported quote that "all wars are planned in boardrooms of [Theftist] banks" reveals such Theftism.
War benefits only the planners and the banks that advance loans to rebuild what they have helped destroy.

After all, "*There are No Enemies Amongst the Dead*"©2019, as explains the title of a poignant poem by Betty Paoli Siegelin, describing the fate of the innocent cannon fodder placed purposely in harm's way by the uncaring sociopaths and psychopaths.

Politics and the Political Party are revealed accurately:

> *"The Party seeks power entirely for its own sake. We are not interested in the good of others; we are interested solely in power, pure power. What pure power means you will understand presently. We are different from the oligarchies of the past in that we know what we are doing. All the others, even those who resembled ourselves, were cowards and hypocrites. The German Nazis and the Russian Communists came very close to us in their methods,*

> *but they never had the courage to recognize their own motives. They pretended, perhaps they even believed, that they had seized power unwillingly and for a limited time, and that just around the corner there lay a paradise where human beings would be free and equal. We are not like that. We know that no one ever seizes power with the intention of relinquishing it. Power is not a means; it is an end. One does not establish a dictatorship in order to safeguard a revolution; one makes the revolution in order to establish the dictatorship. The object of persecution is persecution. The object of torture is torture. The object of power is power. Now you begin to understand me."* — George Orwell, *1984*

A temporary solution: Saving the present American political republic, a short term solution to avoid the current destructions being wrought by the sociopaths and psychopaths working through the centralized political state waste and theft mechanism in Washington D.C. This can and should be constitutionally accomplished via a Convention of States[42]: The ideal temporary solution may be to nullify the District of Columbia and its extraordinary debts, rules, regulations, Laws, etc., and start afresh with just the Property Definition on page 2, above. All the jurisprudence, codes of regulation, taxing authorities, etc. to be abandoned, thus turning Washington D.C. into an exemplar horror museum of what the future of mankind is to avoid. Some military protection will be needed to repel armed political states from expected Theftist attacks. Soon enough, the world's population will see the benefits of Freedom and those armed political states will collapse because they are too expensive to maintain plus history shows that they all eventually collapse.

The inevitable financial collapse of the present dishonest private monopoly Federal Reserve money system will cause a

---

[42] Article V of the United States Constitution.

large temporary destruction of productive economy and culture. The existing “federal reserve notes/deposits” barter facilitating currency will be seen as the gigantic functioning Theftist fraud that it was constructed to be.

A new temporary currency system will arise based upon positive productive value, and not based upon any sort of a monopolistic political decree (a fiat).

A positive productive attitude will become normalized as:

> *“If the creditor has his interest to take care of,*
> *the debtor has his honor to preserve, and the*
> *loss to the one is as fully severe as to the other.”*
> Thomas Paine (1737-1809)

Life is not possible without the defining boundary condition of Property.
Freedom is not possible without Property protection.

Chapter Six

Science, the search for truths of the Universe, utilizing the four step Scientific Method. Property is best built and best protected through the use of science.

1. **Observation** (for data gathering)
2. **Hypothesis Formulation**
3. **Extrapolation** (Extend the hypothesis)
4. **Observation Again** (for corroboration) – Testing.

+ Use **Occam's Razor**[43] for fewest unproven assertions.

Only one failure, on a trip through the Scientific Method, is needed to end a wrong hypothesis. In politics, it takes lies, war, destruction, etc., to make some change.

Choose the Scientific Method *uber alles* for best success in life. After all, Mother Nature relies on reality to make rational sense so you'll be in good company!

Humans have done very well developing science and engineering, through the use of discovery plus invention. The rigorous subject of science underpins all of human life, and is a subject with a long way to advance into the future. Our human destiny is amongst the stars. A first step to fulfill human destiny requires a culture that escapes from the darkness of Theftism's coercion.

One of Theftism's coercions is the lie that science is used by them honestly. Honest corroborating experiments or calculations always expose these lies.

---

[43] After William of Occam (1287 – 1347)

Life is not possible without the defining boundary condition of Property.
Freedom is not possible without Property protection.

Successful Theftists, using Machiavelli's "effective truth" (a lie accepted as truth) never shut up in the face of disproof's of their follies. It is their constant haranguing that may produce an "effective truth" now and again, embodied in some political law or convincing swindle that restricts Human Freedom and therefore the Freedom of other lifeforms.

An example of the misuse of the Scientific Method to promote false concepts:

> "Atmospheric Carbon Dioxide (CO_2) is Bad". To see the truth, all that is needed is to refer to the rigorously inductive science of thermodynamics and the downstream equilibrium equations of physical chemistry. Pioneered by Josiah Willard Gibbs (1839-1903), the statistical mechanics innovation explains possible states of physical systems and the interaction between systems.
> With respect to Carbon, the atmospheric system is in a constant equilibration seeking process with the adjacent plant and ocean systems. The widely advertised political state lie that human-caused CO_2 must be taxed and curtailed is a purposeful worldwide swindle to bypass Freedom and control production of everything. And, to eventually control the numbers of humans exhaling CO_2. Ironically, plant life uses CO_2 to produce Oxygen, O_2, the very gas that animal life needs. More ironic is that those wanting to tax CO_2 are now promoting a plant life diet for corralled humans, a diet that needs more CO_2 to grow extra plants.
> There is no measurable rise in atmospheric temperature due to CO_2 . The opposite is the truth: first the atmospheric temperature changes, then the percentage of CO_2 changes. The percentage of the atmospheric trace gas CO_2 by weight and volume is about a tiny 0.04%. One substance that does vary the atmospheric temperature is visible

water vapor, seen as cloud which is just one of the atmosphere's self-actuating thermostats. Interestingly, sea surface temperatures over the last 3000 years (based on isotope ratios in marine organism remains in the Sargasso Sea) show that the today's average temperature is about the average over the last 3000 years.

Use science properly - There are numerous innovative ways for science and its applications (engineering) to protect property:

1. Production of energy is the primary foundation of a productive society. For example: Small decentralized nuclear power plants to energize a neighborhood of productive humans. These plants would have enough backup capacity to energize portions of other connected neighborhoods if needed.

2. Decentralized power from cosmic energy described by Nikola Tesla. Everything accomplished by Tesla worked very well, so the claims outlined below are very likely to be just as workable.

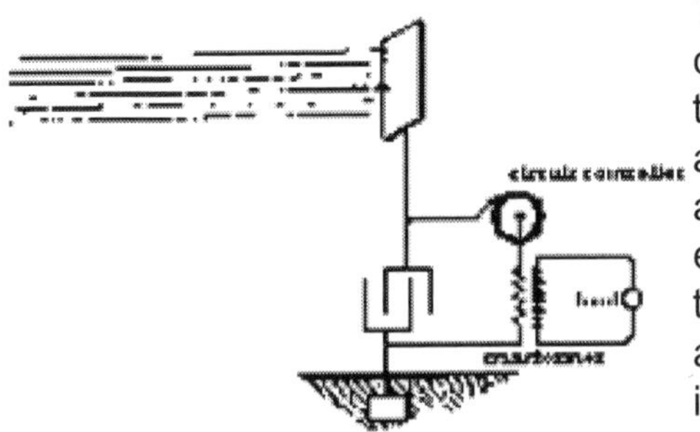

Tesla's intent was to condense the energy trapped between the earth and its upper atmosphere and to transform it into an electric current: A properly tuned system can capture and convert radiant energy in such a prescribed arrangement.

"I have advanced a theory of the cosmic rays and at every step of my investigations I have found it completely justified".

Life is not possible without the defining boundary condition of Property.
Freedom is not possible without Property protection.

"The attractive features of the <u>cosmic rays</u> is their constancy. They shower down on us throughout the whole 24 hours, and if a plant is developed to use their power it will not require devices for storing energy as would be necessary with devices using wind, tide or sunlight".

"All of my investigations seem to point to the conclusion that they are small particles, each carrying so small a charge that we are justified in calling them neutrons. They move with great velocity, exceeding that of light". N. Tesla "*Brooklyn Eagle*" 1932, July 10th

3. Common technologies resulting from science such as locks, confidentiality, contracts/agreements, camouflage, alarms, warning sensors, generation of electrical energy via decentralized methods, etc., are known and available to protect property.

4. Using science, a society of productive humans must rely upon Trust. Trust is achieved through the use of a history of word-keeping, agreement-keeping, and evidence of reliable, durable, and positive productivity. Insurance and savings (of all kinds) are the backstop batteries that save society from serious erosion of trust and Freedom.

5. Other life forms need not be preyed upon by mankind for food. Man can power himself with chemically defined diets such as indicated by "Studies in Metabolic Nutrition Employing Chemically Defined Diets" by Winitz, Seedman, and Graff.

6. Best wishes to all ! This Freedom Building Effort is worthwhile in the extreme

------------end--------------

Made in United States
Orlando, FL
20 August 2022